Naval Surface Fire Support

An Assessment of Requirements

BRADLEY MARTIN, BRITTANY CLAYTON, JONATHAN WELCH,
SEBASTIAN JOON BAE, YERIN KIM, INEZ KHAN,
NATHANIEL EDENFIELD

T0302943

Prepared for the United States Navy
Approved for public release; distribution unlimited

NATIONAL DEFENSE RESEARCH INSTITUTE

For more information on this publication, visit www.rand.org/t/RR4351

Library of Congress Cataloging-in-Publication Data is available for this publication.
ISBN: 978-1-9774-0475-6

Published by the RAND Corporation, Santa Monica, Calif.
© Copyright 2020 RAND Corporation
RAND is a registered trademark.

Cover: U.S. Navy photo by Joshua Adam Nuzzo.

Support RAND
Make a tax-deductible charitable contribution at
www.rand.org/giving/contribute

www.rand.org

Preface

The objective of this study was to develop recommended changes to existing naval surface fire support (NSFS) requirements in light of Chief of Naval Operations and Commandant of the Marine Corps warfighting concepts published since the last NSFS requirements update in 2005, such as Littoral Operations in Contested Environments (LOCE), Distributed Maritime Operations (DMO), and Expeditionary Advanced Base Operations (EABO). This study addresses both the need established for support of forces ashore and the requirements implied for the surface ships that will deliver these effects. This research could form the basis for requirements documents and a capabilities-based assessment, as well as long-term investment.

This research was sponsored by the Office of the Chief of Naval Operations and conducted within the Navy and Marine Forces Center of the RAND National Defense Research Institute, a federally funded research and development center sponsored by the Office of the Secretary of Defense, the Joint Staff, the Unified Combatant Commands, the Navy, the Marine Corps, the defense agencies, and the defense intelligence enterprise.

For more information on the RAND Navy and Marine Forces Center, see www.rand.org/nsrd/nmf or contact the director (contact information is provided on the webpage).

Contents

Figures and Tables

Executive Summary

The objective of this study was to develop recommended changes to existing naval surface fire support (NSFS) requirements. The most recent requirements do not reflect recent operational concepts developed by the U.S. Navy (USN) and U.S. Marine Corps (USMC), such as Distributed Maritime Operations (DMO) and Expeditionary Advanced Base Operations (EABO). But, in fact, the failure to correctly and precisely define requirements has been an issue for decades. Although there have been letters from command in the Marine Corps, the only formal requirements that the Navy uses are decades old and simply reflect a general requirement to perform the mission.

Objectives and Approach

We first examined existing requirements documents and reviewed historical uses of NSFS. This phase also included discussions with NSFS subject matter experts (SMEs) in USN and USMC. We then refined these requirements by applying battalion-level scenarios to support development of baseline fire support requirements for ashore maneuver elements. Next, we applied a formal model to address volume of fire and required magazine size. In parallel, we evaluated the program of record (POR) for NSFS systems and platforms against the projected requirements as derived from a literature review and sponsor-approved scenarios and modeling. We also considered possible technological and platform solutions, although we note that there is not always a formal requirement to guide these efforts. By comparing requirements—

actual and potential—we then made recommendations for requirements and capability development.

Findings

Our research revealed the following findings.

- Targeting, particularly in denied environments, is likely to be
  very challenging and will be highly dependent on organic assets,
  principally unmanned aerial vehicles (UAVs). This issue is not
  confined to NSFS. Targeting in general is likely to be a challenge
  in many warfare areas. However, NSFS requirements for rapid
  and accurate firing information make the problem particularly
  difficult.
- Sensor-to-shooter timelines are far too long to support effective
  engagement on a fluid battlefield. NSFS for maneuvering forces
  ashore has to be capable of responding at very short notice to calls
  for fire.
- A single ship firing rounds from a single gun, even if targeting
  is optimal and command and control (C2) is well executed, is
  physically limited in the targets it can reach and the numbers
  of targets it can simultaneously service. This suggests that a single-ship model simply might be unworkable in heavily contested
  environments, no matter the capability and capacity of individual
  ships. Some autonomous vessel proposals could alleviate some of
  this shortfall, although none of the proposals necessarily result in
  platforms that would actually be capable of sufficient persistence
  and volume.
- The Navy has selected its mix of munitions using considerations
  of what it believes to be its most likely operational mission, but as
  a result has undervalued its magazines munitions that might be
  particularly valuable. In addition, its POR for munitions does not
  address area effects.
- Most of the scenarios we consider involve a considerable expenditure, and formal modeling against a plausible target set indicates
  a very high volume of munitions expenditures, generally beyond

what would be carried in a ship's magazine. Achieving suppression as opposed to destruction lowers demand, as does use of area munitions, but the fundamental conclusion is that although the exact volume requirement is not defined, it is unlikely that the POR can even approach requirements for sustained fire. Some new technologies allowing on-site manufacture of ordnance could mitigate limitations of magazine size.

Recommendations

The most obvious and compelling recommendation is that USMC should identify what it needs from the Navy, using some combination of scenario and quantitative analysis. Absent a formal definition of requirements, the Navy has neither the incentive nor the reason to go beyond what is stated in the ship basis Required Operational Capabilities/Projected Operational Environment (ROC/POE) documents.

- Regardless of what requirements are ultimately ratified, USMC and USN should continue to invest in organic airborne ISR (intelligence, surveillance, and reconnaissance), which can be used even when parts of the electromagnetic spectrum are denied.
- Regardless of eventual requirements determination, USMC and USN should invest in tactical C2 solutions that allow compression of sensor-to-shooter timelines.
- Assuming requirements do get determined according to what seem to be likely scenarios, the following additional investments should be considered:
  - area munitions to challenge enemy maneuver capability
  - lighter munitions that allow extension of range, specifically to allow ships to service multiple landing force targets from a single location
  - ship modifications for larger magazines
  - unmanned fire support platforms that can be put into direct support roles
  - additive manufacturing to allow for production of gun ammunition to increase on-station time during periods of high use.

Acknowledgments

The authors would like to thank those individuals who supported this study. First, we would like to thank MajGen Coffman, U.S. Marine Corps; and MajGen King, Col Thomas Mitalski, and LtCol Ron McLaughlin of the Expeditionary Warfare Directorate of the Office of the Chief of Naval Operations for discussing naval surface fire support requirements with us. We also gratefully acknowledge the helpful advice of CDR Ryan Collins and LCDR Peter Downes of the Surface Warfare Directorate of the Office of the Chief of Naval Operations. Finally, we greatly appreciate the insightful feedback of our reviewers David Luckey and Peter Buryk.

Abbreviations

3D	three-dimensional
A2/AD	anti-access/area-denial
AGS	Advanced Gun System
ANGLICO	Air-Naval Gunfire Liaison Company
ANSF	Advanced Naval Surface Fire
AoA	Analysis of Alternatives
ATGM	anti-tank guided missile
BMG	bulk metallic glass
C2	command and control
CEP	circular error probable
CG MCCDC	Commanding General, Marine Corps Combat Development Command
DDG	Destroyer Designated Guided (guided-missile destroyer)
DMO	Distributed Maritime Operations
DPICM	Dual-Purpose Improved Conventional Munition
EAB	expeditionary advanced base

EABO	Expeditionary Advanced Base Operations
EMRG	Electromagnetic Railgun
GLGP	Gun-Launched Guided Projectile
HE	high explosive
ICD	Initial Capabilities Document
IRGC	Iranian Revolutionary Guard Corps
ISR	intelligence, surveillance, and reconnaissance
JEF	Joint Expeditionary Fires
JFSEOL	Joint Fires in Support of Expeditionary Operations in the Littorals
KPP	key performance parameter
LOCE	Littoral Operations in Contested Environments
LRLAP	Long Range Land Attack Projectile
MCCDC	Marine Corps Combat Development Command
MCTP	Marine Corps Tactical Publication
MEB	Marine Expeditionary Brigade
MEU	Marine Expeditionary Unit
MLRS	multiple-launch rocket systems
NATO	North Atlantic Treaty Organization
nm	nautical mile
NSFS	naval surface fire support
PLAN	People's Liberation Army Navy (China)

PLANMC	People's Liberation Army Navy Marine Corps (China)
POE	Projected Operational Environment
POR	program of record
RAP	rocket-assisted projectile
ROC	Required Operational Capabilities
rpm	rounds per minute
SME	subject matter expert
TF	task force
TLAM	Tomahawk Land Attack Missile
UAV	unmanned aerial vehicle
USMC	U.S. Marine Corps
USN	U.S. Navy

CHAPTER ONE

Introduction

Naval surface fire support (NSFS), referred to in earlier eras as "naval gunfire support," has been a traditional mission of U.S. Navy (USN) surface combatants, and indeed was a major mission envisioned for combatants and as large and emblematic as battleships. Although guns were seen as a major instrument of sea control for the great naval battles that would decide command of the seas, they were also seen, and widely used, as ways to directly influence the battle ashore by providing the equivalent of artillery support for Army and Marine Corps forces operating ashore.

Background

Although much about ground warfare has changed since the last time NSFS was used in decisive ways, fire support certainly has not, and indeed has become the major way of equalizing uneven numbers and using technology to dominate a battlespace. Typically, this fire support refers to air power, but artillery, mechanized and otherwise, was and remains a key component of ground force operations.

USN and U.S. Marine Corps (USMC) have both asserted and shown the value of sea-based fire support options. Indeed, until very recently, the Navy had envisioned the guided-missile destroyer (DDG) 1000 primarily as a ship with signature minimized to allow littoral operations, but with gun and missile systems very directly oriented toward providing fire support for forces ashore.

Although there is no denying that the Navy and Marine Corps have viewed the NSFS mission as important, the actual requirements associated with the mission are broad, to the point that the requirements are, in some cases, no more specific than "include as a ship mission." For that reason, the Expeditionary Warfare Directorate of the Office of the Chief of Naval Operations requested an examination of the formal requirements and programs for NSFS.

Research Approach

We first examined existing requirements documents and reviewed historical uses of NSFS. This phase also included discussions with NSFS subject matter experts (SMEs) in USN and USMC. We then refined these requirements by applying battalion-level scenarios to support the development of baseline fire support requirements for ashore maneuver elements. Next, we applied a formal model to address volume of fire and required magazine size. In parallel, we evaluated the program of record (POR) for NSFS systems and platforms against the projected requirements as derived from a literature review and sponsor-approved scenarios and modeling. We also considered possible technological and platform solutions, although we note that there is not always a formal requirement to guide these efforts. By comparing requirements—actual and potential—we then made recommendations for requirements and capability development.

Report Organization

Six chapters follow this introduction. Chapter Two is a historical overview of combat applications of NSFS, itemization of existing NSFS requirements, and relevant source documents. Chapter Three covers formally stated requirements. Chapter Four presents the results of our scenario analysis and the range of potential fire support requirements derived from the scenario analysis. Chapter Five summarizes the results of volume of fire modeling. Chapter Six describes the NSFS POR and developmental efforts relevant to Chapters Four and Five. Finally, Chapter Seven itemizes our findings and describes recommendations that emerged from this analysis.

CHAPTER TWO

The Historical Context and Current Capabilities

Within the context of amphibious operations, NSFS is and has been typically used as preparatory fire to suppress or neutralize enemy littoral defenses and then to provide fire support for landing forces. The Second World War marked the maturation of NSFS in amphibious operations. Afterward, from the Korean War to modern conflicts, the utility and mission set of NSFS expanded to include defensive operations, interdiction of logistical targets, and strategic messaging. This chapter will focus specifically on the modern history of NSFS and its evolution in terms of capabilities, doctrine, and employment.

Historical Trends and Lessons Learned

When employed effectively, NSFS can provide invaluable firepower to suppress, neutralize, or even overwhelm an adversary. In contested littoral environments, NSFS has historically played a critical role in weakening coastal defenses and providing essential fire support for the follow-on landing forces. In this section, drawing upon historical examples, we examine historical trends and lessons learned from the employment of NSFS and synthesize these trends and lessons for the future evolution of NSFS capabilities.

The Prerequisite of Localized Sea Control

In the past, the utility of NSFS has consistently relied on a fleet's ability to exert or contest localized sea control. To employ naval fires, a fleet must be able to maneuver with relative freedom within littoral zones.

This theme has been demonstrated repeatedly, from the disastrous Gallipoli landings to the successful employment of NSFS in the Pacific campaign during the Second World War. Even during the Vietnam War, vessels in Operation Sea Dragon had to contend with increasingly capable coastal defense batteries along the North Vietnamese shore. Unable to render ineffective enemy batteries to gain localized sea control, U.S. vessels responded by operating at greater stand-off distances and employing maneuver tactics. Historically, submarines, sea mines, and aerial threats have posed the most-significant threats to a fleet executing naval fires.

Now, modern threats include a myriad of missile-based threats. The development of anti-access/area-denial (A2/AD) technology has forced the engagement ranges of NSFS further and further from landing objectives. For instance, "Chinese A2/AD capabilities include long-range precision theater strike systems—primarily cruise and ballistic missiles and manned aircraft—and anti-ship weapons."[1] In particular, anti-ship cruise missiles, such as the Chinese YJ-83 with a range of 160 km, can limit naval maneuverability and ability of ships to employ NSFS.[2] Although state actors possess the most-robust A2/AD technologies, the attack on the USS *Mason* in 2016 reflects the increasing proliferation of such weapon systems to nonstate actors.[3]

Fleets historically have employed three methods in mitigating contested sea control: (1) increasing power and range of naval firepower, (2) isolating the objective through battlefield preparatory actions,[4] and (3) employing overwhelming mass. The Battle of Okinawa serves as a good case study. During the battle, the naval fleet employed highly

---

[1]   Timothy M. Bonds, Joel B. Predd, Timothy R. Heath, Michael S. Chase, Michael Johnson, Michael J. Lostumbo, James Bonomo, Muharrem Mane, and Paul S. Steinberg, *What Role Can Land-Based, Multi-Domain Anti-Access/Area Denial Forces Play in Deterring or Defeating Aggression?* Santa Monica, Calif.: RAND Corporation, RR-1820-A, 2017, p. 75.

[2]   Bonds et al., 2017, p. 76.

[3]   Dan Lamothe, "Missiles from Rebel-Held Yemen Fired at U.S. Navy Destroyer, Saudi Military Base," *Washington Post*, October 10, 2016.

[4]   Battlefield preparatory actions include a range of activities, such as anti-submarine warfare, mine sweeping, aerial interdiction, and deception.

effective naval fires by isolating the island from Japanese naval intervention and endured grueling punishment from kamikaze attacks through sheer naval mass.[5] Although modern threats have intensified and expanded the contest for sea control, naval fires remain relevant and necessary. The aforementioned methods can serve as a foundation for modern naval fires to meet and overcome missile-based threats and contest sea control for the successful employment of naval fires.

The Dilemma of Range, Volume, and Precision

Range, volume, and precision typically serve as the principle metrics of NSFS. Ideally, naval fires would possess abundant range, overwhelming volume, and unprecedented precision. Yet, in reality, naval gunfire historically has been unable to achieve all three characteristics simultaneously. For instance, during the early stages of the Second World War, naval gunfire was executed at long ranges to reduce the fleet's vulnerability. Yet this led to a loss in terms of precision and the volume of fires delivered. Greater stand-off inherently increases flight time of fires and subsequently undermines volume of fires. Conversely, later in the war, ships sacrificed range by closing the distance to their targets to achieve gains in volume and precision.

The tradeoff between range, volume, and precision persisted during the Vietnam War. Because of political and humanitarian constraints, greater precision was necessary to avoid civilian casualties. For example, while providing fire support on an enemy assault at Mo Duc, the USS *Stormes* maneuvered closer to the target before firing on the village. Yet, even when range and volume were sacrificed, the level of precision required to target an enemy located in civilian villages proved difficult to achieve. The USS *Stormes* repelled the assault but killed numerous civilians. Although the Navy could sacrifice one attribute to achieve gains in the other attributes, the possible gains were limited.

This trend has held true into the modern era with the advent of advanced munitions and missile technology. Although Tomahawk Land Attack Missiles (TLAMs) and advanced precision munitions,

---

[5]   Jeter A. Isely and Philip A. Crowl, *The U.S. Marines and Amphibious War: Its Theory, and Its Practice in the Pacific*, Princeton: Princeton University Press, 1951, pp. 520–563.

such as the Long Range Land Attack Projectile (LRLAP), provide remarkable range and precision, the cost of such munitions makes achieving the necessary volume cost-prohibitive. For instance, during Operation Odyssey Dawn in Libya in 2011, the cost of munitions, which primarily consisted of TLAMs, totaled $340 million.[6] The termination of the LRLAP for the Advanced Gun System (AGS) of the Zumwalt-class destroyer also serves as an example.[7] The dilemma of range, volume, and precision will almost certainly remain a key challenge in the evolution of NSFS into the future. Most likely, there will be no easy, simple solution to this dilemma. As in the past, the solution will involve multiple platforms, tactical adaptions, and organizational reforms.

The Necessity of Munition Diversification
As previously discussed, NSFS is typically quantified in terms of range, volume, and precision. However, the types of munitions fired are equally as important as the guns that launch them. During the First World War, the inefficiency of NSFS was caused largely by prevalence of armor-piercing rounds with delayed fuses. Originally designed for penetrating the armor of ships, the armor-piercing rounds performed terribly in land bombardment. They often buried themselves deep into the ground before exploding—typically producing a thunderous roar but little actual damage.[8]

In comparison, during the Second World War, the evolution of naval munitions played a significant role in the success of NSFS. For instance, white phosphorous projectiles were used to flush Japanese defenders from their covered positions. Five-inch star shells, which provided intense illumination, proved crucial in assisting ground forces in repelling Japanese counterattacks. The demand for 5-inch star shells

---

[6]   Jeremiah Gertler, *Operation Odyssey Dawn (Libya): Background and Issues for Congress*, Washington, D.C.: Congressional Research Service, March 30, 2011, p. 22.

[7]   U.S. Government Accountability Office (GAO), *Weapon Systems Annual Assessment: Limited Use of Knowledge-Based Practices Continues to Undercut DOD's Investments*, GAO-19-336SP, Washington, D.C., May 2019, p. 106.

[8]   Isely and Crowl, 1951, pp. 38–39.

was so intense that the supply never met the demand. Meanwhile, high-velocity rounds from battleships were employed against concrete pillboxes.[9] The variety of munitions during this period enabled NSFS to provide a variety of effects on the battlefield.

The basic types of NSFS munitions have not changed since the Second World War, principally relying on high explosives and illumination. NSFS lacks the variety of munitions available to its sister function of field artillery.[10] Presently, the U.S. Army is in the process of upgrading and modernizing artillery long-range fires. The upgrade program includes efforts to improve power generation, increase range and rate of fire, and acquire different types of munitions.[11] Similarly, if NSFS is to evolve into the future, NSFS munitions must develop to include a wide gamut of functions, including smoke and submunitions. Thus, NSFS should not be measured solely by range, volume, and precision—but also by the diverse set of effects NSFS can employ on the modern battlefield.

The Value of Training and Specialized Units
Throughout the history of NSFS, its success relied heavily on well-trained, specialized units. As NSFS fire missions and tasks grew more complicated, close coordination and specialized units were increasingly necessary. Two notable examples are riverine units in Vietnam and Joint Assault Signal Company/Air-Naval Gunfire Liaison Company (ANGLICO) units from the Second World War to the present.

During the Vietnam War, USN created two riverine units, the River Patrol Force and the Mobile Riverine Force, both equipped with specialized NSFS capabilities that were vital to the success of U.S. units operating along the Mekong River. With little riverine experience since

---

[9] Donald M. Weller, "The Development of Naval Gunfire Support in World War Two," in Merrill Bartlett, ed., *Assault from the Sea: Essays on the History of Amphibious Warfare*, Annapolis, Md.: Naval Institute Press, 1983, pp. 276.

[10] Brian Duplessis, "Thunder from the Sea: Naval Surface Fire Support," *Fires*, May-June 2018, p. 54.

[11] Todd South, "Plan Would Double Artillery Upgrades in Army Arsenal over the Next Five Years," *Army Times*, March 22, 2019.

the American Civil War, USN created specialized vessels to handle the shallow waters and close-quarters combat necessitated by the environment. Likewise, USN created specialized command and communication structures, allowing Army units assaulting the shore to rapidly call in and receive NSFS. During the Tet Offensive, the riverine NSFS vessels were invaluable in blunting the North Vietnamese offensive. The performance of riverine NSFS vessels during Tet were not an isolated occurrence. Rather, these specialized capabilities were able to provide critical close fire support deep within enemy territory throughout the duration of the war.[12]

At present, ANGLICO units provide the ability to coordinate and employ fires from air, land, and maritime assets. Yet the mission of ANGLICO units—unlike their predecessors—has expanded and evolved in conjunction with wider Marine Corps concepts, such as Expeditionary Force 21.

Current Capability to Meet Naval Surface Fire Support Requirements

There is no specifically labeled NSFS POR aimed at satisfying NSFS requirements. However, the requirement to provide NSFS is of long standing and is primarily associated with ship missions rather than as separate capability requirements. There is a Required Operational Capabilities/Projected Operational Environment (ROC/POE) requirement to deliver fire support, and the Navy equips its ships with guns and ammunition to do this, specifically the 5-inch gun mounted on the DDG 51–class destroyers and the CG 47–class cruisers.[13] The DDG 1000 class, which was originally intended as a land-attack destroyer with significant NSFS capability, went through a series of

---

[12] John Darrell Sherwood, *War in the Shallows: U.S. Navy Coastal and Riverine Warfare in Vietnam, 1965–1968*, Washington, D.C.: Naval History and Heritage Command, Department of the Navy, 2015, pp. 319–320.

[13] Chief of Naval Operations, Surface Warfare (N96), *Required Operational Capabilities and Projected Operational Environment for (Arleigh Burke) Class Guided Missile Destroyers*, OPNAV F3501.311C, July 28, 2017.

delays and cost escalation, ultimately resulting in truncation of the class to three ships and re-designation of its role as anti-surface warfare.[14]

Five-Inch Gun

The Navy's Mk 45 5-inch gun, shown in Figure 2.1, is a fully automatic gun mount used primarily for NSFS. It was initially deployed in 1971 and replaced the legacy Mk 42 5-inch/54-caliber gun. The Mk 45 was designed to be lighter in weight and easier to maintain than its predecessor. The gun mount includes a 20-round automatic loader drum with a maximum firing rate of 16 to 20 rounds per minute (rpm). It

Figure 2.1
Mk 45 5-Inch Gun

SOURCE: USN.

---

[14] O'Rourke, Ronald, *Navy DDG-51 and DDG-1000 Destroyer Programs: Background and Issues for Congress*, Washington, D.C.: Congressional Research Service, April 16, 2018.

can be operated by the Mk 160 Gun Computer System or the Mk 86 Gun Fire Control System.[15]

There are three variants of the Mk 45 currently being used in the fleet: Mods 1, 2, and 4. Mk 45 Mods 1 and 2 are 54 caliber with a barrel length of 270 inches. Mk 45 Mod 4 includes several improvements to previous mods, including a longer barrel of 310 inches (62 caliber). The Mk 45 Mod 4 also improves gun performance and maintainability.[16] Mk 45 Mods 1, 2, and 4 are employed on the *Arleigh Burke*–class destroyers (DDG 51) and the *Ticonderoga*-class cruisers (CG 47). An estimated inventory of available Mk 45 gun mounts is shown in Table 2.1.

Table 2.1
Mk 45 Inventory

Ship Class	Hull Numbers	Number of Hulls	Mk 45 Mod 1	Mk 45 Mod 2	Mk 45 Mod 4
CG 47 class (two mounts/ship)	CGs 61, 65–68	5	10	N/A	N/A
	CGs 69–73	5	N/A	10	N/A
	CGs 52–60, 62–64	12	N/A	N/A	24
DDG 51 class (one mount/ship)	DDGs 51–80	30	N/A	30	N/A
	DDGs 81–117 (and forward)	37	N/A	N/A	37
Estimated inventory		**89**	**10**	**40**	**61**

SOURCES: U.S. Navy, "Fact File: Cruisers – CG," webpage, last updated January 9, 2017; U.S. Navy, "Fact File: Destroyers – DDG," webpage, last updated August 21, 2019b.

NOTE: N/A = not applicable.

---

[15]   U.S. Navy, "Fact File: MK 45 - 5-Inch 54/62 Caliber Guns," webpage, last updated January 16, 2019a.

[16]   U.S. Department of the Navy, *U.S. Navy Program Guide 2017*, Washington, D.C., undated.

DDG 51 Class

The DDG 51–class guided missile destroyers, also known as the *Arleigh Burke*–class destroyers, are multi-mission warships capable of conducting antiair warfare, anti-submarine warfare, and anti-surface warfare. The lead hull was delivered in 1991, and a total of 67 hulls have been delivered as of June 2019 (DDG 117 was delivered on June 7, 2019).[17] The DDG 51 class has been produced in three different "flights," which refer to different capability packages. Hulls continue to be procured, with an additional 21 hulls authorized to be constructed.[18] Each hull of the DDG 51 class accommodates one Mk 45 gun mount with a magazine capacity of 600 rounds. Thirty hulls host the Mk 45 Mod 2 variant, while 37 host Mod 4.

CG 47 Class

The CG 47–class guided missile cruiser, also known as the *Ticonderoga*-class cruiser, shown in Figure 2.2, was the first ship class to host the Aegis combat system. The lead hull was delivered in 1983, with an additional 26 hulls delivered through 1994. The first five hulls, however, were too expensive to modernize and were removed from service.[19] Thus, 22 hulls remain active. Each hull accommodates two Mk 45 gun systems, with a total magazine capacity of 1,200 rounds. Five hulls host the Mk 45 Mod 1 variant, five host the Mod 2 variant, and 12 host the Mod 4 variant.

Naval Surface Fire Support Current Capability Shortfalls

The challenges of a littoral environment reinforce the need for sufficient range of NSFS systems. With a maximum navigational draft between 31 and 33 ft,[20] the DDG 51 and CG 47 hulls are not able to get very close to shore. (The five-inch gun, on the other hand, has a maximum range of 13 miles.) Littoral areas are often congested and

---

[17] Naval Vessel Register, "USS Paul Ignatius (DDG 117)," webpage, June 7, 2019.

[18] Naval Vessel Register, "No Name (DDG 138)," webpage, September 27, 2018.

[19] O'Rourke, 2018.

[20] Naval Vessel Register, "USS Arleigh Burke (DDG 51)," webpage, June 27, 2018; Naval Vessel Register, "TICONDEROGA (CG 47)," webpage, December 12, 2017.

Figure 2.2
CG 47–Class Cruiser

SOURCE: USN.

not easily navigable. With these factors in mind, meeting the range requirement for the NSFS mission is paramount.

The volume of fire requirement, while not clearly defined, is likely not met with the current NSFS arsenal. DDG 51 has a rate of 20 rpm. With its 600-round magazine, that would allow 60 minutes of continuous suppression. Likewise, the CG 47 has a rate of 20 rpm (with two 5-inch guns mounted). With its 1,200-round magazine, it can also achieve 60 minutes of continuous suppression. Although this requirement is not explicitly quantified, 60 minutes of suppression likely does not meet the need for full NSFS support.

DDG 51 and CG 47 are multi-mission surface warships that serve the Navy in many areas of combat. As a result, these vessels are in high demand, which can lead to a lack of availability when required for NSFS. In concert with this, these ships have multiple ammunition requirements that result in limited onboard magazine storage space. With limited storage and the high rate of fire required to achieve

NSFS, these ships must be resupplied frequently. Resupply can be accomplished at sea via underway replenishment or by transiting to an available port. Either method requires resources and interrupts mission accomplishment. Considering all of these factors—in particular, the high demand of the platforms and their frequent need to resupply—availability of NSFS platforms presents a challenge.

DDG 1000 was intended as a ship platform specifically to provide fire support, whose Operational Requirements Document (ORD) values included an AGS firing LRLAP ordnance capabilities. However, because of the expense of the ship and changes in the mission requirement, the ship class was cut from 32 hulls to its current three. The costs of rounds were greatly increased, and the Navy lost the expected economy of scale benefits of the order of LRLAP rounds, resulting in a November 2018 decision to place the AGS guns in layup. At the same time, the mission of the DDG 1000 class was shifted from land attack to offensive surface strike. The Navy accordingly revised the DDG 1000 ORD and requested from the Joint Requirements Oversight Council (JROC) key performance parameter (KPP) relief for the AGS. ORD revision 2 was approved by the Joint Staff J-8 chaired Functional Capabilities Board in February 2019. The result is that the Navy's NSFS capabilities still reside entirely in the DDG 51 and CG 47 classes.

CHAPTER THREE

Naval Surface Fire Support Capability Development Requirements

Ship ROC/POE guidance specifies that a ship must be capable of performing a mission, but generally does not specify mission requirements to a high level of detail. These requirements are covered in detailed mission area documents. Unlike other maritime requirements, the Navy's NSFS mission requirements lie at the intersection of both the naval and the landing force; thus, the Navy cannot singularly generate meaningful requirements. The requirements instead reside with the landing force and derive from what that commander might require for fire support in the absence of adequate organic capability. The initial NSFS capability gaps were outlined by USMC in a 1992 Mission Needs Statement and have been updated in a series of letters from senior USMC leaders to Navy staff. The documents that more formally translate requests into requirements have not progressed to the point of mandating specific performance parameters.

The "Hanlon Letter"

USMC has made landing force requirements for NSFS known through a series of official letters from the Commanding General, Marine Corps Combat Development Command (CG MCCDC) to Navy staff. The first such letter was sent in 1996, with a follow-on letter in

1999.[1] The most detailed letter was sent in 2002 with the subject line "Naval Surface Fire Support Requirements for Expeditionary Maneuver Warfare."[2] This letter, signed by LtGen Edward Hanlon, Jr., is often referred to as the "Hanlon letter" or the "2002 letter." It noted, "Over the past . . . [six] years, this Command produced . . . [letters], outlining the Marine Corps' requirements for Naval Surface Fire Support (NSFS). As we progress in this critical area of force protection and expeditionary littoral warfare, we find it necessary to emphasize and further clarify our NSFS requirements."

These letters emphasized range requirements consistent with ranges envisioned for the USMC concepts of Expeditionary Maneuver Warfare (EMW) and Operational Maneuver from the Sea (OMFTS). The 2002 CG MCCDC letter provided detailed threshold and objective requirements for seven capabilities: system response, range, accuracy and precision, target acquisition, ordnance effects, volume of fire, and sustainment. This letter also established the requirement that NSFS should provide each landing infantry battalion with fire support equivalent to that provided by each battalion's direct support 155-mm artillery battery. This concept is sometimes referred to by Marines as the "155-mm battery equivalency," and it was widely supported throughout our research and SME discussions. There was a qualifier occasionally added to the 155-mm battery equivalency, which was that it needed to have the range to support the vertical assault element of the landing force, a range capability not currently possessed by a USMC 155-mm battery.

---

[1]   Shawn A. Welch, *Joint and Interdependent Requirements: A Case Study in Solving the Naval Surface Fire Support Capabilities Gap*, master's thesis, Norfolk, Va.: Joint Advanced Warfighting School, 2007.

[2]   Edward Hanlon, Jr., "Naval Surface Fire Support Requirements for Expeditionary Maneuver Warfare," memorandum to Chief of Naval Operations (N7), Quantico, Va., March 19, 2002. (Note: The CG MCCDC has been dual-hatted as a deputy commandant; currently, the title is Deputy Commandant for Combat Development and Integration.)

Definitions Used in the Letter Drove Requirements Specification
The following definitions, used in the 2002 CG MCCDC letter,[3] were drawn from the fire support doctrine of the time.

- Suppression: "Fires on or about a weapons system to degrade its performance below the level needed to fulfill its mission objectives, during the conduct of the fire mission" (Marine Corps Tactical Publication [MCTP] 3-10F);[4] "Temporary or transient degradation . . . below the level needed to fulfill . . . [the] mission objectives . . . [of a weapon system]" (Joint Publication [JP] 3-01).[5]
- Neutralization: "Fire which is delivered to render the target ineffective or unusable" (Allied Administrative Publication [AAP] 06);[6] "Fire which is delivered to hamper and interrupt movement and/or the firing of weapons" (MCTP 3-10F).[7]
- Destruction: "Fire delivered for the sole purpose of destroying material objects" (MCTP 3-10F).[8]

MCTP 3-10F and the 2002 CG MCCDC letter reference joint doctrine that eliminated these terms in 2014. This further complicates a discussion of NSFS requirements, since 2002 joint doctrine has evolved and does not currently provide sufficient definitions of such effects as suppress, neutralize, and destroy.

Given the difficulty of quantifying some of the requirements for the seven capabilities outlined in the 2002 CG MCCDC letter, Navy program managers tended to focus on the one capability requirement, range, that had definitive mid-term threshold (63 nautical miles [nm]) and objective numerical goals (97 nm). These relatively long ranges were driven by USMC's assessment of the threat to NSFS ships while

---

[3] Hanlon, 2002.

[4] USMC, *Fire Support Coordination in the Ground Combat Element*, MCTP 3-10F, Washington, D.C., April 4, 2018.

[5] Chairman of the Joint Chiefs of Staff, *Countering Air and Missile Threats*, JP 3-01, Washington, D.C.: Joint Staff, April 21, 2017.

[6] North Atlantic Treaty Organization, *NATO Glossary of Terms and Definitions (English and French)*, AAP-06, Brussels, Belgium, 2019.

[7] USMC, 2018.

[8] USMC, 2018.

USMC was developing a new vertical assault aircraft, the MV-22, which had a "combat radius" or round-trip ship-to-shore range of 400 nm.[9] The 2002 CG MCCDC letter stated: "The mid-term objective range requirement for naval guns was calculated by adding the operational radius of the current medium-lift assault support system to the maximum range of the most commonly fielded threat fire support system."[10] Many SMEs felt that USMC stating a range requirement that presumed the NSFS platform's location at sea instead of stating needed effects relative to the landing force location ashore also drove some Navy staff officers to focus on range requirements over options for getting close enough to deliver desired effects regardless of range. Table 3.1 summarizes the parameters identified in the letters, which, again, can only be viewed as desired characteristics, not validated requirements.

The gaps are summarized in the table. What is noteworthy is that we are nearly at the end of the periods identified in the letter, and effectively none of the requirements have been achieved.

The 155-mm Equivalency

Because the 155-mm battery equivalency was so widely supported as a requirements metric, the research team reached out to the Marine Corps Combat Development Command (MCCDC) for the basic *day of ammunition* (DoA), or combat load, for the average 155-mm battery. The USMC planning factor for a 155-mm battery is 720 rounds broken out by type, as shown in Table 3.2. The planning factors are subject to local standard operating procedures, operation planning, commander's guidance, etc. Also note that the planning factors continue to assume that a Dual-Purpose Improved Conventional Munition (DPICM) replacement will be fielded, although, to date, one has not, which often results in DPICM being replaced by high explosive (HE), bringing the total number of HE rounds to 600. The planning factors also assume that all vehicles and personnel organic to the battery are available to move this quantity of ammunition.

---

9   Boeing, "V-22 Osprey," webpage, undated.

10   Hanlon, 2002.

Table 3.1
Naval Surface Fire Support Requirements Summary Matrix

Characteristic	Key Performance Parameter	Near Term (2004–2005)	Middle Term (2006–2009)	Far Term (2010–2019)
System response	Threshold	2.5 minutes	2.5 minutes	2.5 minutes
	Objective	Limits of technology	Limits of technology	Limits of technology
Range: naval guns	Threshold[a]	41 nm	63 nm	97 nm
	Objective	63 nm	97 nm	Limits of technology
Range: other systems	Threshold	200 nm	200 nm	262 nm
	Objective	222 nm	222 nm	Limits of technology
Accuracy and precision	Threshold	50-m CEP[b]	50-m CEP	50-m CEP
	Objective	20-m CEP	20-m CEP	20-m CEP
Target acquisition	Threshold	50 nm	63 nm	97 nm
	Objective	63 nm	97 nm	Limits of technology
Ordnance effects	• Ordnance effects destroy or suppress point, area, and moving targets, including personnel and material, and destroy hardened targets. • Ordnance effects provide smoke, illumination, and incendiary effects.			
Volume of fire	• Volume of fire is equally important as precision. • Volume is needed for mass fires, suppression, combined arms effects, and close fire support. • Sufficient quantities are maintained to sustain desired effects over time.			
Sustainment	• All systems are sustainable via underway replenishment.			

SOURCES: Hanlon, 2002; GAO, *Defense Acquisitions: Challenges Remain in Developing Capabilities for Naval Surface Fire Support*, Washington, D.C.: GAO-07-115, November 2006.

[a] *Threshold* is the minimal acceptable value to meet the KPP; *objective* is the maximum desired requirement.

[b] CEP = *circular error probable*, which measures precision and is defined as the radius of a circle in which 50 percent of the fired rounds are expected to land.

Table 3.2
155-mm Battery Day of Ammunition

Type of Projectile	Number Required
Projectile, 155-mm Smoke White Phosphorus XM1121	24
Projectile, 155-mm Smoke White Phosphorus M825	40
Projectile, 155-mm Illuminating VL M1124	30
Projectile, 155-mm Illuminating IR M1123	10
Projectile, 155-mm High Explosive Rocket-Assisted M549A1	16
Projectile, 155-mm HE (IMX-101) with Supplemental Charge	408
Projectile, 155-mm Excalibur Increment 1-B	0
Projectile, 155-mm Alternate Warhead Cluster Munition	192

SOURCE: MCCDC, Fires and Maneuver Integration Division, *Naval Surface Fire Support Requirements and Program Updates*, information paper, Quantico, Va., May 1, 2018.

Documents Since the Hanlon Letter Have Addressed Gaps but Not Specified Requirements

The Joint Fires in Support of Expeditionary Operations in the Littorals (JFSEOL) Initial Capabilities Document (ICD), Joint Expeditionary Fires (JEF) Analysis of Alternatives (AoA), and draft Advanced Naval Surface Fire (ANSF) ICD represent efforts to take the "small r" requirements of the CG MCCDC letters and develop "capital R" requirements in Service and Joint documentation processes. We list these documents and their implications in chronological order:
- In 2005, the Joint Requirements Oversight Council validated the JFSEOL ICD, which noted that fires "includes the triad of fires delivered from aircraft, ships/submarines and ground assets."[11]

---

[11] Joint Requirements Oversight Council, *Initial Capabilities Document for Joint Fires in Support of Expeditionary Operations in the Littorals*, Washington, D.C.: U.S. Joint Chiefs of

The JFSEOL ICD highlighted four NSFS capability gaps: command and control (C2) systems, target engagement, accuracy and precision, and volume of fires. Detailed requirements to quantify volume and suppression, however, were not addressed in significant detail. Furthermore, the JFSEOL ICD did not quantify range; rather, it used such wording as "extended range munition choices" would be required for future fires capabilities.

- In 2009, the JEF AoA was published, highlighting the importance of NSFS engaging moving targets in all weather, conducting fires in proximity to friendly forces, and providing volume effects (i.e., suppression).[12]
- In 2016, an ANSF capabilities-based assessment was conducted, resulting in a 2017 ANSF ICD. Although the ANSF ICD remains in draft form, it identifies five prioritized NSFS gaps, largely consistent with the 2005 JFSEOL ICD. It focuses on range, target engagement, counterfire, volume of fires, and accuracy and precision:[13]
  - Range: Engage fixed targets at sufficient range and effectiveness to support operations ashore (near-term interim threshold of 41 nm; far-term threshold of 97 nm).
  - Target Engagement: Engage moving/relocatable point and area targets under restricted weather conditions.
  - Counterfire: Provide counterfire capabilities against rockets, artillery, mortars, and missiles.
  - Volume of fires: Provide fires to achieve volume effects against surface targets.
  - Accuracy and precision: Engage targets when friendly forces are in close contact or collateral damage is a concern.

These gaps are not specific in terms of measurable output, and they are framed as areas where improvement is required.

Staff, JROCM 274-05, November 1, 2005 (summarized in GAO, 2006).

[12] Center for Naval Analyses, *Naval Surface Fire Support AoA Final Report*, Arlington, Va., November 2009.

[13] MCCDC, 2018.

Formal Requirements Summary

In summary, the existing NSFS requirements literature is robust, but much of it is in the form of free text letters and not the more-commonly accepted Joint Capabilities Integration and Development System (JCIDS) requirements documents. Furthermore, the free text letters often went beyond desired effects and delved into what tactics, techniques, and procedures the NSFS platforms would be using to deliver these effects, often assuming a position relative to units ashore that resulted in greatly increased range requirements. As the Navy and Marine Corps begin to move Distributed Maritime Operations (DMO), Expeditionary Advanced Base Operations (EABO), and Littoral Operations in Contested Environments (LOCE) from concepts to capabilities, they should examine options for defining NSFS requirements in terms of desired effects and, in turn, ensuring that those desired effects are documented in JCIDS requirements documents. This research can form the basis for requirements documents and a capabilities-based assessment, as well as long-term investment.

CHAPTER FOUR

Scenario Analysis as a Basis for Additional Requirements

The current formal assessment of NSFS requirements does not appear to reflect warfighting requirements, which leaves the dilemma of what the correct set of requirements might be. Requirements are developed relative to a likely set of missions and threats. These missions and threats can be aligned to a set of scenarios, which can represent the interaction between missions, threats, and capabilities.

The study team was able to attend a wargame out-brief and an operational advisory group session to obtain added comments on NSFS requirements.

- At the 21st Century Fires #2 Main Event Wargame out-brief, the study team noted that all three teams in the wargame emphasized "target acquisition, long-range ISR [intelligence, surveillance, and reconnaissance] resulting in targetable data, and volume fires." Two teams noted the requirement for NSFS ships to provide "artillery-like" support and that NSFS platforms will need to be maneuverable and able to defend themselves.
- At the Marine Expeditionary Unit (MEU) Operational Advisory Group, the study team heard from several MEU and Amphibious Squadron commanders who agreed that NSFS should provide 155-mm battery equivalency but added the qualifier that it needed to have the range to support the vertical assault element of the landing force in the event that the MEU's "long-range artillery" (AH-1Z Super Cobra and F-35B Lightning II) was not available. No specific range was discussed, other than 155-mm range

beyond the lead element of the landing force and the comment, similar to that of the wargame out-brief, that NSFS platforms will need to be able to defend themselves to move in close enough to provide such support.

Although NSFS has historically been employed to suppress or provide cover for maneuvering elements transitioning from ship to shore, DMO, LOCE, and EABO may require sustained NSFS support during subsequent operations ashore by the landing force. Detailed requirements for DMO, LOCE, and EABO have not yet been developed. This study addresses both the need established for support of forces ashore and the requirements implied for the surface ships that will deliver these effects.

Four Scenarios for Analysis

The following scenarios were selected to represent reasonable examples where NSFS is likely to be needed. They are neither interrelated nor designed to take place concurrently. Each scenario focuses on supporting a Marine infantry battalion or equivalent-sized task force. In two cases, the unit ashore is part of a Marine Expeditionary Brigade (MEB), and in the other two, it is part of an MEU. This choice reflects the assumption that a guided-missile destroyer could approximate the firepower of a USMC artillery battery—the default unit required to support an infantry battalion. All scenarios deliberately exclude air support and the sufficient ground-based fires to isolate what exclusive reliance upon NSFS would require. Each scenario possesses unique attributes designed to examine general rather than system-level fire support requirements and considerations.

These scenarios were selected after discussion with the sponsor and stakeholders and were selected for several reasons. The first is that they represent plausible cases, sometimes closely aligned with actual occurrences, in which NSFS could be essential capability. These are more than theoretically possible and are, in fact, based on situations where NSFS was needed and/or desirable. The scenarios were based both on historical use and the inputs of stakeholders. They are also aligned with the scenarios that the Navy and Marine Corps are considering in development of new operational concepts. These scenarios stress several different capabilities inherent in NSFS, such as range, volume, and accuracy. Effective use in scenarios such as these indicates likely value in capability development. These scenarios are not being offered as definitive requirements, but as illustrative cases. If these scenarios are used in formal requirements documents, there would have to be an additional step of validation by the joint and intelligence communities. However, they are robust, plausible, and illustrative.

In any scenario, ship numbers will be limited and the ships are subject to tasking for multiple missions. Because we are attempting to represent the landing force requirements—and not necessarily the Navy's ability or inability to meet those requirements—we will highlight the requirements of the landing force as opposed to the supporting ship's availability or capability. These are critically important, but the first step is to describe the range of capabilities potentially needed.

It is important to note that these scenarios are representative and should not be considered as authoritative planning guidance. By design, however, they are plausible and provide a general guide as to what landing force commanders might require. We will match this with more-formal modeling efforts to address the range of NSFS requirements.

Scenario 1: Palawan, Philippines

In the first scenario, shown in Figure 4.1, USMC MEB is in the process of deploying across the Pacific theater in response to a sudden significant increase in regional aggression by China. The first unit to arrive is a Marine infantry battalion, which distributes itself across the province of Palawan in several expeditionary advanced bases (EABs). The battalion's main effort has been tasked with achieving sea control of the Mindoro Strait, at the northern end of the province. Its primary supporting effort has been tasked with achieving sea control of the Sulu Sea, at the southern end of the province. The People's Liberation Army Navy (PLAN) has recognized the risk posed by USMC control of Palawan and has deployed an amphibious demonstration force that includes a PLAN Marine Corps (PLANMC) brigade (approximately 3,000 marines) embarked on several amphibious transport docks.

Figure 4.1
Palawan Scenario

When the PLAN amphibious task force gets close to Palawan, it splits in two and deploys PLANMC forces to seize key terrain, at the northern and southern straits—the same locations as the USMC EABs.

When the PLANMC forces land and move inland, III MEB (forward) leadership fears the EABs are at risk of being overrun by larger and better-supported forces. The PLANMC has come ashore with ZTD-05 amphibious assault vehicles, ZBD-05 amphibious infantry fighting vehicles, and PLZ-07 (Type 07B) amphibious 122-mm self-propelled howitzers.[1] The battalion's mortars are spread out between

---

[1]   IHS Markit, "Jane's Sentinel Security Assessment–China and Northeast Asia," webpage, February 13, 2019. See also: Office of the Secretary of Defense, *Annual Report to Congress: Military and Security Developments Involving the People's Republic of China 2018*, Washington, D.C.: U.S. Department of Defense, May 16, 2018.

the EABs and add little value given the threat, and the battalion does not yet possess artillery support ashore. F-35s are in Japan to provide combat air patrols.

Target Acquisition

Target acquisition is first achieved by RQ-21 Blackjacks, which had been providing ISR. However, shortly after contact begins, they are jammed and prove no longer viable. Ground-based USMC observers visually acquire targets and submit subsequent calls for NSFS. The MEB (forward) fire support coordination center requests NSFS approval from the Joint Task Force, who promptly grants it.

Fire Support

USMC observers identify multiple hasty targets and request simultaneous engagement to maximize effects. USMC EABs, including the observers, begin experiencing communication challenges, believed to be the result of PLAN jamming.

Challenges

Several significant challenges exist for effective NSFS within the Palawan scenario. Ideally, the response time for fire missions would be roughly two and half minutes from call for fire to rounds out and roughly ten minutes until first impact. The scenario is constructed assuming a need for support in multiple locations, separated by distances greater than the range of the supporting ship's guns. To be close enough to service one target set, the ship will be separated from the competing request. Thus, within this scenario, additional gun range would be advantageous. A related issue is that with a single gun, even for targets within range, there are limitations on the number of targets that the gun can effectively service. If the enemy attacks nearly simultaneously at multiple points, the ship simply will not be able to meet multiple target sets at a time.

The scenario posits a contested electromagnetic environment and subsequent challenges overcoming disruptions in targeting and command and control. With the landing force dispersed in a fluid environment, knowing where to fire quickly and accurately is essential but is likely to pose a major challenge. This particular scenario suggests that shortfalls in ISR and C2 may be particularly important to overcome in meeting NSFS requirements.

Volume would not necessarily be advantageous if the major challenge is hitting the correct target in a dense environment. However, if volume coverage is desired, the ship possesses a limited magazine and does not possess area munitions—such as the DPICM. As a result, the ship's fire support would be limited to the delivery of HE munitions at limited range and in limited quantity.

Scenario 2: Aden, Yemen

In the second scenario (shown in Figures 4.2, 4.3, and 4.4) in response to a severe earthquake, an Amphibious Ready Group with an embarked MEU is ordered to the Arabian Sea to execute a humanitarian assistance/disaster relief mission in Aden, Yemen. The initial force deployed ashore consists of a battalion-sized task force (TF) composed of forces

Figure 4.2
Aden, Yemen Scenario

Figure 4.3
Yemen Scenario, Detailed Airfield Graphic

from the MEU's Ground Combat Element and Logistics Combat Elements. The TF establishes itself at Aden International Airport, just north of the city.

Not long after establishing its position, the TF is attacked by several seemingly coordinated suicide small unmanned aircraft systems (sUASs), or drones, and suffers a missile impact. The attacks are believed to have originated from Al Anad Air Base. The base was recently retaken by the Houthis (also known as Ansar Allah), who are believed to be accompanied by Iranian Revolutionary Guard Corps (IRGC) advisors. Their suicide sUAS systems and tactics, techniques, and procedures are widely known. The missile is believed to be an SS-21 Scarab-A (Tochka), a Battlefield Short-Range Ballistic Missile.[2]

---

[2]   Missile Defense Advocacy Alliance, "OTR-21 Tochka (SS-21 Scarab)," webpage, 2019.

Figure 4.4
Yemen Scenario, Al Anad Air Base

The TF commander realizes that the force needs immediate fire support to counter the threat. The Aviation Combat Element's fire support resources, including assigned F-35s, are unavailable. The MEU artillery battery and its M777 155-mm howitzers were not part of the initial offload. Additionally, the adversary position is beyond the range of the MEU's organic 81-mm and 120-mm mortars. As a result, the TF requests NSFS to neutralize the threat posed by adversary fires until an organic fire capability can be established by rotary-wing or ground-based fires.

Target Acquisition
The MEU had already deployed an RQ-21 Blackjack to survey the damage from the earthquake. It is promptly re-tasked with conducting surveillance and target acquisition over Al Anad Air Base and confirms that six SS-21 Scarabs are located on the north side of the runway.

One appears to have just fired a missile, and the others remain loaded. Additionally, there are an estimated 200 to 300 military-aged males in mixed military uniforms with military equipment, apparently preparing for a convoy.

Fire Support
The TF calls for immediate suppression of the SS-21s to ensure that they are unable to launch additional missiles. The TF also requests the neutralization of the convoy, which is believed to be preparing to maneuver toward the TF's position at Aden International Airport. The Marines require close supporting fires because they are focused on the most immediate threat. The purpose of the fire support is to neutralize the adversary's missiles and ground forces.

Challenges
Within this scenario, several challenges exist for the effective employment of NSFS. Given the mobile and multiple nature of the missile threat, target acquisition will rely heavily on forward observers, ground-based unmanned aerial systems, and efficient coordination between the supported ground unit and the supporting naval vessel. The ship possesses no organic systems that significantly assist with target acquisition.[3] In the Palawan scenario, simple geographic dispersion of the enemy force posed the principal challenge. This scenario, by contrast, highlights the challenge of highly mobile and dispersed enemy forces. The ship's single 5-inch gun simply cannot respond quickly enough to provide support against multiple moving targets.

This scenario does not pose the denied electronic environment challenges evident in the Palawan scenario. It does, however, have the same challenges associated with volume and persistence of fires. Ships inherently possess limited magazines and do not possess munitions appropriate for area fires, such as DPICM. Even with effective targeting and favorable engagement geometry, limited magazine capacity could make meeting requirements challenging.

---

[3]   Hanlon, 2002, p. 9.

Scenario 3: Latakia, Syria

In the third scenario, shown in Figure 4.5, following the conclusion of the Syrian civil war, a USMC infantry battalion is deployed in Latakia, Syria as part of a UN-sanctioned peacekeeping force. The battalion's task was to assist in maintaining a truce in the city of Latakia and the surrounding region. However, the truce did not endure, and, exploiting civil strife, Hezbollah fighters have established a foothold within the city and on its outskirts. To undermine the fragile truce, the rebel force has begun employing rocket and missile fire against U.S. peacekeeping forces.

Equipped and advised by the IRGC, Hezbollah forces employ both rockets and anti-tank guided missiles (ATGMs) from multiple urban firing positions. According to intelligence reports, the opposing force consists of two company-sized elements, divided between well-concealed positions within the city and well-defended firing positions

Figure 4.5
Latakia Scenario

on the outskirts of the city. Within the city, using commercial drones for ISR and targeting, Hezbollah fighters employ a combination of small arms (AK-47s), heavy machine guns (DShK), anti-tank rockets (AT-4), and RPG-29s. Leveraging the presence of civilians and the urban environment to their advantage, they employ rapid, complex hit-and-run tactics on U.S. convoys.

Outside the city, another Hezbollah force accompanied by IRGC advisors is providing long-range fire support. Most of its rocket force consists of several Katyusha multiple-launch rocket systems (MLRS), a 122-mm 9M22 Grad model. The Hezbollah rocket force is concentrating its fires on identified U.S. military positions and the city's port. At the same time, sustained rocket and missile fire severely limit the maneuverability of the Marine battalion. The consistent barrage of rocket fire is inflicting considerable damage to the port, undermining the ability of the Marine peacekeeping force to receive supplies or reinforcements. There are also unverified reports of a C-802 Noor variant, an anti-ship missile, operated by Iranian military personnel in conjunction with Hezbollah forces. With a range of roughly 120 km, the C-802 Noor variant could pose significant risk to surface ships.

The Marine battalion subsequently requests NSFS to neutralize adversary firing positions, both within and outside the city. Foremost, the Marines require the immediate and imminent threat of Hezbollah rockets to be nullified. At present, the Marines lack the range to target and render combat ineffective the well-entrenched positions outside the city. While the threat of Hezbollah rockets persists, the Marines will be unable to maneuver freely and eliminate the Hezbollah fighters within the city.

Target Acquisition

Upon receiving fire, ground-based RQ-21 Blackjacks are launched, mainly to acquire targets. The firing capabilities available are limited. The battalion's organic mortars lack sufficient range to target the defensive positions outside the city. Additionally, given the nature of their mission, U.S. forces do not have attached artillery, and there are no close air support assets available. As a result, fire support requests are sent directly to the nearby surface vessel to conduct NSFS.

Fire Support

Within the scenario, the nature of the fires emphasizes close support-ing fires, as they are focused on the immediate adversary threat. Given their range and lethality, the primary goal of supporting fires is to neutralize the threat of Hezbollah rocket fire. Emphasizing immedi-ate and rapid close supporting fires, the supporting vessel will strive to suppress the rocket positions or neutralize the Katyusha MLRS. How-ever, the fortified nature of the Hezbollah rocket positions will most likely render the naval fires ineffective. Meanwhile, firing into a dense urban environment to support Marines within the city will require a high-degree of precision. Hezbollah forces will also quickly target U.S. unmanned aerial vehicles (UAVs) to undermine the sensor-to-shooter chain, further complicating the targeting challenge.

Challenges

This scenario is similar in several important respects to the Aden sce-nario, in that the mobility of the rocket systems poses a challenge for targeting and timely engagement. With a single surface ship in sup-port, the ship will have to cycle through the fire missions in sequen-tial order. Each mobile target will require roughly six rounds to create a standard 200-meter effect area.[4] The reinforced defensive positions of the adversary forces further complicate the situation. Lacking any munitions other than HE, the ship's guns will likely be ineffective against well-fortified Hezbollah defensive positions.

The urban environment highlights some of the inherent limi-tations of NSFS. Even if sensor-to-shooter timelines could be greatly compressed, simply finding the right targets in an urban environment can be challenging. For instance, buildings and the presence of civil-ians complicate firing geometries and targeting. Moreover, no matter how accurate and timely targeting might be, NSFS will likely inflict collateral damage.

---

[4]   Hanlon, 2002, p. 16.

Scenario 4: Tallinn, Estonia

In the fourth scenario, shown in Figure 4.6, MEB is in the process of compositing and deploying to Norway as part of a U.S. European Command flexible deterrent option. The first Marine infantry battalion to arrive in theater was pushed forward to support the United Kingdom–led North Atlantic Treaty Organization (NATO) Enhanced Forward Presence battalion in Tapa, Estonia. The remainder of the MEB is steadily flowing into Norway and unable to provide support at this moment.

While transitioning ashore at Tallinn, U.S. and NATO forces are attacked by a company-plus-sized element of Russian special forces disguised as proxy nationalist militias. Using indirect fires, such as rockets and mortars, Russian special forces attempt to harass and delay U.S. forces. To deny or disrupt access to the strategic port of Tallinn, Russian special forces seek to fix U.S. forces as they transition ashore. By

Figure 4.6
Tallinn Scenario

using a combination of rocket and mortar fire, Russian forces intend to force U.S. forces to retreat from the port or sufficiently delay the build-up of U.S. forces until reinforcements arrive from the Russian Western Military District.

The Russian force is equipped with mortar systems (2B14 Podnos 82-mm mortar system), rocket-propelled grenade launchers (RPG-28; RPG-30), ATGMs (9M133 Kornet), man-portable air-defense systems (9K333 Verba), and small arms and heavy machine guns. The Russian special forces are supported by a small detachment of rocket artillery using the BM-21 Grad (122-mm multiple rocket launcher). The threat environment is further complicated by Russian army-level and brigade-level capabilities, such as surface-to-air missiles (SAMs), and Russian combat aviation assets, such as attack aircraft, rotary-wing aircraft, and bombers. The presence and range of Russian integrated air defense systems and SAMs, such as the S-400 and S-300, pose the most significant risk to U.S. and NATO air assets and ground forces. Meanwhile, submarines of the Baltic Fleet also pose a substantial risk to U.S. and NATO surface vessels operating in theater.

As a result, the Marine battalion requests NSFS to neutralize or destroy adversary indirect fire assets and delay the adversary with interdicting fire on main avenues of approach. The goal is to neutralize high-value Russian rocket systems and delay Russian consolidation of forces. With ample naval fires, the Marine battalion intends to build combat power ashore, focusing on transitioning its M777s ashore. Understanding the risks to naval vessels, the Marine battalion will have to maneuver quickly, exploiting the brief window of opportunity.

Target Acquisition

When Russian proxy forces open fire on U.S. forces, several airborne RQ-21 Blackjacks identify adversary positions. UAVs identify four different target groups, ranging from two to six vehicles each, consisting of 2S23 Nona-SVKs (120-mm self-propelled mortar system) and BM-21 Grads. Upon target acquisition, targeting information is relayed to the shipborne fire cell for firing solutions. As new targets appear, forward observers relay targeting data to the supporting surface vessel.

Fire Support

Upon receipt of target information from ground-based UAVs, the primary focus will be the neutralization of Russian rocket artillery. Given that Russian units will be using a "shoot-and-scoot" tactic, the time between target identification and a firing solution will be condensed. Given the mobility of the targets, speed will be a key factor. The secondary focus will be to delay Russian forces from advancing on main avenues of approach. This will allow U.S. and NATO forces to constitute their defenses and combat power while simultaneously limiting the adversary's mobility and maneuverability.

Challenges

This scenario presents several challenges for effective NSFS. In terms of fires, the scenario highlights the challenge of targeting and successfully neutralizing mobile targets, such as the BM-21 Grads and 2S23 Nona-SVKs. For mobile targets, the sensor-to-shooter chain must be shortened and accelerated, particularly if HE rounds are being relied upon. Area-effect munitions, if such are available, could help limit adversary mobility. However, as noted in the other case studies, ship magazines are largely limited to HE. Finally, the limited magazine size and the resulting inability to sustain fires for an extended period is a significant limitation. In this scenario, the landing force is likely to require near-continuous fire support, and a guided-missile destroyer magazine can service only about an hour of continuous fire. Even if naval fires are measured in its response, a single DDG 51 would still be challenged to meet the demands of the supported ground elements in a prolonged engagement before requiring resupply.

Overall Scenario Findings

The scenarios highlight several challenges associated with employing naval guns to support landing force operations.
- Targeting: Although UAVs can provide useful, real-time data, locating an enemy that is determined either to deny the electro-

magnetic spectrum or to use concealment is a major challenge for NSFS.

- Sensor-to-shooter timelines: NSFS for maneuvering forces ashore has to be capable of responding at very short notice to calls for fire. In addition, it has to be capable of engaging multiple targets. Shortfalls in both capabilities came up multiple times in both scenarios that involved maneuvering forces ashore.

- Simultaneity: A guided-missile destroyer with only one gun is unable to engage multiple targets at the same time. This is particularly problematic when there is a requirement for immediate suppression against multiple firing points—a likely adversary tactic.

- Range: Although there is no stand-off distance sufficient to put ships outside the range of land-based threats, lack of range exacerbates inability to engage dispersed targets. In the cases described here, a ship might not be able to support more than a single landing force element at a time, in large part because of the range limitations of shipborne guns. Thus, the effective area that a single naval vessel can support is constrained to a few miles.

- Munitions types and load-out: HE destruction rounds are essentially the only munition carried in Navy cruisers' and destroyers' gun magazines. However, the scenarios repeatedly demonstrate that the most-effective munitions for the missions described are combinations of DPICM and HE. Singular reliance on HE is, in fact, exacerbating issues of insufficient volume. A single DPICM round might, in fact, perform suppression and mobility disruption more effectively than multiple HE rounds.

- Volume and persistence: A ship can loiter and remain ready to fire for a long time, but it can only briefly sustain a high volume of fire, measured at most in hours. In some of the cases described here, having high-intensity fire available for only very limited periods either imposes risk on the landing force or requires a considerable force structure of ships ready to relieve one another on the gun line. Larger magazines are likely to be a significantly more efficient way of meeting the volume and persistence requirements.

We will examine more-formal methods of determining volume requirements in the next chapter.

Scenario Implications for Program Execution

Scenarios as we used them are at most heuristic tools, and they are certainly not the basis for requirements determination or PORs. However, they do identify gaps in capabilities which, taken together, bring into serious doubt the Navy's ability to meet what would appear to be reasonable landing force requirements—such as the ability to deliver fire for more than an hour. That is, admittedly, assuming a need for near-continuous high-volume fire, but it does not seem reasonable or responsible to simply discount it.

CHAPTER FIVE

Volume Requirements Modeling

The scenarios represent an attempt to show the most-important requirements and gaps in NSFS, given a set of possible operational contexts.
Many of these gaps already have readily applicable modeling and technical analysis that could be used as a basis for KPPs in systems development. For example, if range or rapidity of target selection and assign-
ment appear as the most-pressing problems that the scenarios identify,
there would be ways to add range or compress the target cycle, all of
which include readily measurable variables.

As noted in the scenario discussions, not all scenarios are likely
to require a high volume of fire. If the engagements are taking place
in urban environments where there is an interest in minimizing collateral damage, volume is less important than precision. Similarly, if the
engagements take place at a distance from the battlefield, a few rounds
with an extended range might be preferable to a large number that
cannot reach the target. However, while volume and persistence—the
application of volume across time—are identified as shortfalls, there
is no model set being used to assess the number of rounds that might
be needed in these likely scenarios. This chapter discusses an approach
that allows quantitative assessments of actual rounds needed in combat
scenarios.

The model depictions are highly generalized and, to be actually
used in a requirement, would have to be paired with a set of scenarios
with agreed-upon threats and constraints. However, they do give an
idea of how many rounds would be required in the scenarios identi-

fied. This, in turn, bears on required magazine size and replenishment capability.

Basic Modeling Parameters for Volume Requirements

To define the volume of fire requirement, the research team conducted modeling scenarios under various conditions. Using a Monte Carlo simulation developed in Python, we tested NSFS volume requirements for the following scenarios:

- random distribution of targets
- denser target groupings
- hardened targets
- suppression fire.

Basic parameters for all models included the following: The field was constructed as a 16 (4 x 4)–mi^2 grid. The field units of the model were in terms of square feet. We assumed one gun on one ship, with a constant firing rate of 20 rpm. The gun was simulated as pointing at one target per shot, moving from target to target after destroying each. Its blast radius was set to 164 x 164 ft^2 (50 x 50 m^2). The CEP for the gun was set to 656 ft (200 m). This meant that the actual locations of the shots differed from where the gun set its target because of imprecision. Unless otherwise stated, simulations were run until at least 30 percent of targets were destroyed completely. We quantified this as destruction. In addition, 6,000 targets were initially placed, unless otherwise stated. A target was considered destroyed once it had been hit twice within a blast radius. Targets were randomly dispersed across the field. One hundred runs were conducted for each scenario, with the exception of 1,000 runs for suppression fire.

We also validated the results from these Monte Carlo simulations using a similar RAND modeling platform called CATAPULTA, or Covert and Aerial Threat Analysis Program to Understand the Lethality of Targeting of Airbases. CATAPULTA is a program written in Microsoft Excel Visual Basic for Applications (VBA) that also uses Monte Carlo simulation to analyze interactions between projectile munitions and targets. It was adapted to meet NSFS specifications.

These models are highly idealized and should be considered representative rather than determinative of volume requirements. However, various elements of the model can be varied to portray different results. For example, any area of desired neutralization or suppression could be selected, and different areas would yield different results. This model should be viewed as a tool and not as a determinative combat simulation.

Random Distribution of Targets

Most models for targeting effectiveness assume suppression is equivalent to 10 percent of targets eliminated, neutralization is 30 percent of targets eliminated, and destruction is 80 percent of targets eliminated. According to Alexander, 1977, "Neutralization of a target is achieved when 30% of the personnel and material are rendered ineffective or 30% of the target area is damaged."[1]

In the first modeling scenario, targets are randomly placed in the open, and the simulation is run until 30 percent of targets are destroyed. The average total number of rounds fired to neutralize 30 percent of targets was 4,800, and it took an average of 240 minutes. On average, 29 shots were required to destroy a single target. The results are presented in Table 5.1. Although this is a simplified scenario, it demonstrates that the inaccuracy of current NSFS munitions requires a large volume of fire to make up for loss of precision. Neutralization also requires more munitions than can be carried by a single guided-missile cruiser or destroyer without needing resupply. Although different ship employment considerations could mitigate this problem, it nevertheless demonstrates that a large volume of fire is required to neutralize targets.

---

[1]   Robert Michael Alexander, *An Analysis of Aggregated Effectiveness for Indirect Artillery Fire on Fixed Targets*, thesis, Atlanta, Ga.: Georgia Institute of Technology, 1977.

Denser Target Groupings

For the second modeling scenario, we kept many of the same parameters from the first scenario but changed the density of the target groupings. This scenario allowed us to test the change in required volume of fire if the targets were closer together, thus mitigating some of the inaccuracy of the munitions. Instead of having 6,000 randomly placed targets, we divided the targets into three groups of 2,000. The adjustment parameter was the radius of the area the targets were placed within. Tables 5.2 and 5.3 display the results for two different target densities. These scenarios demonstrate that fewer munitions are required because damage is also inflicted on unintended targets. Although this allows NSFS ships to meet volume requirements with current inventories, the unintended casualties inflicted would be detrimental in areas where noncombatants are present.

Hardened Targets

To add complexity to the base scenario and to reflect other target types, we explored the requirement for volume of fire to destroy personnel in hardened bunkers. Bunkers were set to be a 12-x-12–ft² area. Because a bunker is a hardened target, we increased the number of hits required to destroy a bunker area to 10, and each bunker held one person. We ran simulations for one, two, and three bunkers randomly placed in a field. The results are shown in Tables 5.4 and 5.5.

We also ran a variation where we had denser bunker placement. In this scenario, 20 bunkers were randomly placed within a 400-by-400–ft² area in the field. The results are presented in Table 5.6.

The hardened targets scenario showed that destroying hardened targets requires, on average, twice the number of munitions.

Table 5.1
Random Distribution of Targets, Modeling Results

Mean Rounds	SD Rounds	Mean Time (Minutes)	SD Time (Minutes)	Mean Targets Left	SD Targets Left	Mean Hits per Target	SD Hits per Target
4,804.25	124.28	240.21	6.21	4,199.76	0.49	28.75	27.23

NOTE: SD = standard deviation.

Table 5.2
Denser Target Groupings, Modeling Results

Radius (ft)	Mean Rounds	SD Rounds	Mean Time (Minutes)	SD Time (Minutes)	Mean Targets Hit	SD Targets Hit	Mean Hits per Target	SD Hits per Target
1,000	198.55	31.98	9.93	1.6	4,188.51	9.23	0.03	1.14
2,000	566.96	78.99	28.35	3.95	4,195.6	3.97	0.09	2.14

NOTE: SD = standard deviation.

Table 5.3
Ninety-Five-Percent Confidence Intervals for Denser Target Groupings (Rounds)

Radius (ft)	Lower	Upper
1,000	192.282	204.818
2,000	551.578	582.442

Table 5.4
Hardened Targets, Modeling Results

Number of Bunkers	Mean Rounds	SD Rounds	Mean Time (Minutes)	SD Time (Minutes)	Mean Targets Left	SD Targets Left	Mean Shots per Target	SD Shots per Target
1	219.67	64.17	10.98	3.21	0	0	223.07	63.67
2	400.10	107.96	20.01	5.40	0	0	212.35	59.42
3	591.34	168.30	29.57	8.41	0	0	216.77	57.95

NOTE: SD = standard deviation.

Table 5.5
Ninety-Five-Percent Confidence Interval Bounds for Denser Bunker Scenario (Rounds)

Bunker	Lower	Upper
1	207.09	232.25
2	378.94	421.26
3	558.35	624.33

Suppression Fire

The three previous modeling scenarios focused on neutralizing targets. In our fourth scenario, we tested what the volume of fire requirement would be to achieve desired suppressive effects. For this excursion, we continued with the denser target placing (three groups of 2,000) from the second model, but modeled *suppression*, previously defined as 10 percent of targets being destroyed (rather than 30 percent).[2] Our interest was in determining how long suppression can be maintained until 600 rounds—the capacity of a guided-missile destroyer—have been depleted. Three hundred people were randomly distributed in a 1-x-1–mi^2 grid.[3] Because suppressive fires merely prevent the enemy from conducting its mission, but fail to eliminate targets, we repopulated the field over set time increments to simulate fire team–sized units being able to return to the fight. In this model, four targets were randomly repopulated back to the field every 30 seconds once suppression had been achieved. Assuming there is one guided-missile destroyer with 600 rounds, suppression fire could be maintained for approximately 40 minutes before the ship would need to replenish or be relieved by another asset. The results are shown in Table 5.7.

Modeling Findings

The use of generalized models helps further define the volume requirement for NSFS under four different conditions. The models find that
- the high CEP and small target size contribute to a requirement for higher volume of fire to achieve desired effects
- most scenarios require more munitions than a single guided-missile cruiser or destroyer has the capacity to carry
- although denser groupings resulted in lower numbers of required munitions to achieve desired effects, the model demonstrates that in scenarios with noncombatants present, the inaccuracy of the munitions will cause significant damage to unintended targets
- destroying hardened targets (bunkers) requires, on average, twice the number of munitions

Table 5.6
Denser Hardened Targets, Modeling Results

Mean Rounds	SD Rounds	Mean Time (Minutes)	SD Time (Minutes)	Mean Targets Left	SD Targets Left	Mean Shots per Target	SD Shots per Target
328.93	76.71	16.45	3.84	0	0	35.50	72.70

NOTE: SD = standard deviation.

Table 5.7
Suppression Fire Modeling, Results

Mean Rounds	SD Rounds	Mean Time (Minutes)	SD Time (Minutes)	Mean Targets Left	SD Targets Left	Mean Hits per Target	SD Hits per Target
607.43	4.92	39.15	1.29	271.25	1.41	23.38	26.86

NOTE: SD = standard deviation.

- suppressive fire on a company-sized element is sustainable for 40 minutes before resupply or relief from another asset is required.

The overall suggestion is that magazine capacity is going to be insufficient for any scenario where there is a serious need to destroy targets or where the suppression requirements are large. The Navy could generate nothing like the NSFS capability and capacity that it used to support landings of the size or type carried out during World War II, which included the ability to attack hardened bunkers and provide sustained suppressive fires. Although the ability to support this kind of operation has not been identified as a requirement, it is important to highlight that this historical usage simply could not be replicated with the magazine capacity present today or contemplated in future combat systems.

The model is intended as a tool that allows planners and ship system requirements developers to understand the tradeoffs available to support the NSFS mission. Although the model is highly abstracted, the message is clear: If high volume and persistent fire support is an objective, ship capacity is not sufficient to meet the objective.

CHAPTER SIX

Naval Surface Fire Support Developmental Efforts

The POR capability for NSFS remains the 5-inch guns associated with cruiser and destroyer platforms. Our review of scenarios and modeling of volume requirements indicate that this is not sufficient, although the exact degree to which it would be sufficient depends on better definition of the formal requirements. We did, as part of the study, examine some possible technological and program options to meet possible requirements, and these are examined below.

Some kinds of capabilities are common to a variety of missions and probably should be developed irrespective of the NSFS mission. These include improved organic ISR—largely provided by airborne platforms—and improved C2 systems. Other such capabilities include improved data analytics and autonomous systems. However, these kinds of capabilities are not unique to NSFS and thus are likely to be developed in any case. The capabilities discussed below are unique to addressing NSFS gaps.

We also do not consider in detail alternative firing platforms. Although it is possible to mount guns or missiles on ships other than cruiser-destroyer platforms, all lack some essential feature for NSFS, most commonly capacity to hold more than a few rounds. Some larger platforms—such as amphibious ships—could be modified for large capacity, but they are in even shorter supply than cruisers and destroyers, with many competing missions. As we have seen, most of the requirements for NSFS have been associated with ship classes, and there have not been capabilities developed that are independent of the host platforms. What we will discuss here largely applies to improving the capabilities of systems already on a ship class, as dictated by ROC/POE.

Capabilities for Range Improvement

Finding ways to extend range has been one place where the Navy has attempted to improve capability with several developmental programs. The history of the AGS and its associated LRLAP has been discussed in the POR section. The requirements associated with the gun and projectile were tied to the ship, and the Navy has changed the ship's mission and requested relief from the range requirement. There have, however, been other developmental efforts.

Rocket-Assisted Projectile

Rocket-assisted projectiles (RAPs) are ammunition that incorporates a rocket motor for independent propulsion. The Navy LRLAP projectile, which we have discussed, is rocket-assisted, but it has been cancelled because of prohibitive cost. Other kinds of RAPs are being developed by the Army, with the HERA and XM1113 systems. These projectiles could help increase ranges of existing weapon systems, but all involve high production costs and would require modification of shipboard magazines to accommodate storage of the highly explosive shells. The high production and ship modification costs appear to make these choices undesirable for use on Navy ships delivering NSFS.

The Navy's Electromagnetic Railgun

The *electromagnetic railgun* (EMRG) is a gun that uses electromagnetic force from the ship's power supply to fire a projectile. It has been in development since 2005 and was originally planned to be an NSFS weapon. Two prototypes have been procured to date, one by BAE Systems and another by General Atomics Aeronautical Systems. With possible range falling between 50 and 100 nm,[1] the EMRG could possibly satisfy the range requirement defined in the Hanlon letter. The EMRG program, however, faces many challenges. As currently designed, EMRG has a low firing rate. The goal is to achieve ten shots per minute, but only a rate of a few shots per hour has been achieved to date. With considerable wear and tear from the speed and heat of the projectile, the launcher barrel has a limited barrel life. Also, because the launcher leverages electrical power from the ship, integration with an appropriate platform remains a challenge.

Hypervelocity Projectile

Throughout development of the EMRG, the Navy recognized its projectile's compatibility with other weapon systems. This projectile, known as the *Gun-Launched Guided Projectile* (GLGP) or the *Hypervelocity Projectile*, can be fired from a 5-inch gun (mounted on the DDG 51– and CG 47–class hulls) or a 155-mm gun (mounted on the DDG 1000–class hulls). If used with the 5-inch gun, the GLGP achieves a range of 26 to 41 nm.[2] With the 155-mm gun, the GLGP achieves a range of 40 nm. Although this does not achieve the threshold range of 63 nm, it is an improvement from the current POR. Additionally, the GLGP is much more affordable than the LRLAP. With its modular design, the projectile presents an opportunity for savings because it can be configured for multiple gun systems. Leveraging systems that are already fielded would enable rapid deployment of the projectile, making it available more quickly than a new design. The GLGP might present a quickly fielded alternative for improving performance of the NSFS mission.

Assessment of Range-Improvement Technologies

The Navy has tried in developmental efforts to meet the general range requirements outlined in the letters sent from USMC. However, these efforts have not transitioned to a POR, and all face technical or storage

---

[1]   Ronald O'Rourke, *Navy Lasers, Railgun, and Hypervelocity Projectile: Background and Issues for Congress*, Washington, D.C.: Congressional Research Service, October 21, 2016.

[2]   O'Rourke, 2016.

issues. In the case of the EMRG, the low firing rate might in fact run directly contrary to requirements for volume and persistence. There might be a case for including the GLGP in the Navy POR as a means of improving range, and we know from the scenario analysis that ability to reach multiple aim points is an important capability, even accepting that it will be impossible to completely move outside enemy coastal defense engagement range. We have not conducted a detailed program analysis to determine whether this system is the best addition to the Navy's POR. We do note that this system would help with at least one capability gap.

Materials to Improve Projectile and Barrel Performance

These materials might have several potential applications. They might conceivably allow construction of lighter projectiles, which could increase range without adding explosive weight or rocket propulsion. They might also strengthen gun barrels, allowing a higher firing rate. The materials discussed below have not been specifically developed for NSFS applications, but they have been used successfully in similar applications.

Shape Memory Materials

Shape memory materials are materials that have memory and thus revert to their original form when heated after deformation. Some newer materials also respond to a magnetic field instead of temperature. Of available shape memory alloys, nickel titanium (also known as *nitinol*) is best researched and is already in use in biomedical and engineering applications.

Because of these materials' ability to retain their initial shape at high temperatures, they might prove to be an improved projectile material that retains shape over a longer range. Possible pitfalls include high cost of manufacture and fatigue failures, although with ammunition, there would be no need for repeated phase changes.

Graphene

Graphene is a two-dimensional form of carbon where the carbon atoms are arranged in a hexagonal lattice. Its two-dimensional structure makes it uniquely strong and electrically and thermally conductive.

Research shows that in proportion to its thickness, graphene is about 100 times stronger than steel. Graphene could be used to construct longer gun barrels for longer range. It could also be used to coat the inside of gun barrels to more efficiently disperse heat and thus increase the rate of fire.

Although the ceiling for innovative uses of graphene is high, graphene faces several setbacks in practical implementation. It is rather brittle and lacks toughness. Additionally, scientists have yet to discover a cost-effective method of large-scale manufacture. Lastly, there are difficulties in translating an inherently two-dimensional material into three-dimensional applications. These difficulties are somewhat alleviated by ongoing research into composite materials, which help plug weaknesses in materials while preserving their strengths.

Self-Healing Materials

Self-healing materials are artificial materials that automatically repair damage to themselves without external human intervention. They are available in all classes of materials (polymers, metals, ceramics, and cement), although most research has been conducted on polymers. Often, these materials are seeded with a self-healing agent that helps fill in gaps.

Possible NSFS applications include coating the inside of gun barrels, which would enable a faster rate of fire because of the material's ability to mend damage caused by rapid firing. Additionally, these materials could reduce maintenance-related needs by reversing the effects of wear and tear on equipment. Such a reduction could free up valuable space on the ship that could be used to store additional ammunition.

Challenges in implementing self-healing materials include their difficult, material-specific fabrication. Additionally, depending on the specific material, healing could be triggered by certain conditions (for example, temperature or heat thresholds), and extensive healing could lead to depletion of healing agents.

Bulk Metallic Glasses/Amorphous Metals

Bulk metallic glasses (BMGs) are solid metallic materials with an amorphous crystal structure. Most metals naturally form crystal grains at the atomic level. Because ordinary metals tend to break along grain boundaries, amorphous metals are stronger and tougher than those with grains. BMGs also have great thermal formability because of the lack of change in volume between molding and cooling.

Because of their superior strength and wear resistance, BMGs could be used to coat gun barrels and thus improve rate of fire. Additionally, their high elasticity and strength-to-weight ratio could help projectiles travel farther, increasing effective range. In the past, BMGs have been investigated for their use in kinetic energy penetrators and as casings for lightweight fragmentation bombs.[3]

Although the lack of grains in BMGs does increase strength, toughness, and elasticity, it also makes the material more brittle and prone to catastrophic failure. Additionally, BMGs tend to revert to their crystalline state given time and higher temperatures. They are also limited by size and cost in manufacture. However, there is research into the creation of BMG-based composite materials, which could solve some of these problems.

Assessment of New Materials in Naval Surface Fire Support Applications

New materials might help with lighter projectiles and thus increase range, but their primary possible application is in improving gun barrels and thus allowing more-prolonged use, possibly with a higher rate of fire. This could help with issues of persistence and volume, but the primary limitations on these capabilities lie primarily with available magazine capacity rather than with limitations on the gun barrel's firing life. As with other capabilities noted in this section, the exact

---

[3] Mark Telford, "The Case for Bulk Metallic Glass," *Materials Today*, Vol. 7, No. 3, March 2004, pp. 36–43.

value of this kind of technology is difficult to assess absent a definitive statement of the requirement.

Technologies to Improve Magazine Capacity and Volume

Ships are designed with a particular magazine size, which is intended to be adequate for project missions. As we have seen in both the scenarios and the volume analysis, ability to stay on station and deliver fires is a major limitation, due largely to the inadequate number of rounds available relative to the requirement. The technologies considered here are designed to allow either more-efficient use of available space or more-ready manufacture of ammunition.

Cluster Munitions

Cluster munitions are weapons that release a large number of smaller submunitions. Cluster munitions have been in consistent use internationally since World War II.[4] They can be both air-dropped or ground-launched and remain relatively inexpensive compared with firing an equivalently large number of separate rounds or missiles. Thus, they are a cost-effective way to raise volume of fire without needing to deploy additional artillery systems. They are also a mature technology, needing no additional development to be fielded.

Cluster munitions do suffer from less-precise targeting because of their wide area of effect, which can lead to greater likelihood of hitting civilian targets. Additionally, submunitions frequently do not explode immediately and lack a self-destruction capability, producing "duds" that can explode years later, after cessation of conflict. This dud rate remains as high as 20 percent, despite a decade-long U.S. initiative to lower that rate to less than 1 percent.

The humanitarian aspects of using such weapons cannot be ignored. However, neither can their potential usefulness in the kinds of stressful scenarios we have examined. In cases where high volume

---

[4] Cluster Munition Coalition, "Use of Cluster Bombs," webpage, undated. (Note that this source is a nonprofit opposed to the use of cluster munitions.)

is necessary, the area to be affected is large and the targets numerous, with a high likelihood of collateral damage regardless of the munition used. Having these kinds of ordnance available would add flexibility in cases where volume is the requirement and collateral duty less of a consideration.

Loitering Munitions

Loitering munitions are autonomous UAVs that loiter in a given area until they can home in on a detected threat. Although this does not in itself add volume or capacity, it does potentially improve the chances of hitting a target using fewer munitions. Rather than firing several rounds at dispersed targets, the loitering munitions attack targets upon recognition of specific threat parameters. The munitions use electro-optical, radio, microwave, or infrared frequency sensors to detect threats.

These UAVs can fly out beyond safe range for sea-based platforms to provide additional fire support. As a benchmark, the IAI (Israel Aerospace Industries) Harpy has a 100-km range and can loiter for up to two hours.[5] Loitering munitions are cheap and accurate because of their smaller size and on-system camera guidance. These vehicles are also hard to track and kill because of their low radar, visual, and thermal signatures. Additionally, they are already in use in the United States and several other countries, most prominently Israel, and thus do not need additional development to field.

Possible concerns include the range and flight time of the UAV. Rather than fly at the speed of gun ammunition, loitering munitions fly at normal UAV speeds once launched. If in orbit over a target area, they can hit a target rapidly. If the call for fire is in an unexpected area, however, it will take time for the UAV to reach the target. Another concern is that the UAV is larger than normal gun ammunition and will take up space on the ship. The tradeoff might be, then, that it takes fewer rounds to destroy a target, but fewer rounds are available to be carried.

---

5   Yaakov Lappin, "IAI Announces New Mini Harpy Loitering Munition," *Jane's Defence Weekly,* February 19, 2019.

Additive Manufacturing

Three-dimensional (3D) printing is the fabrication of physical objects from a digital file. It could help with the issue of insufficient magazine capacity by allowing local fabrication of gun ammunition. *Additive manufacturing* is the primary technique used today to three-dimensionally print objects, where the object is built layer by layer. 3D printing allows for rapid prototyping of designs and reduces the need for long supply chains because products can be printed on the spot. 3D printing also accommodates complex geometries and is highly customizable.

Because of the decentralized nature of production and the ability of one printing machine to produce multiple designs, 3D printing could potentially increase efficiency in magazine capacity by producing projectiles on board naval platforms. For example, the GLGP uses sabots to fit each propulsion system. With a 3D printer, sabots for the appropriate system could be printed on demand, instead of predetermining stocks of each system. Additionally, platforms might be able to store base materials for parts instead of spare parts used in repair and maintenance, freeing further space needed for magazines. The use of additive manufacturing can also reduce waste material because objects are built layer-by-layer instead of by carving away excess materials.

Recent success stories in industrial applications of 3D-printed parts support the feasibility of such applications. In 2015, the new Airbus A350 XWB consisted of more than 1,000 3D-printed parts.[6] Earlier that year, the Royal Air Force Eurofighter Typhoon also flew with 3D-printed replacement parts.[7] In August 2013, NASA successfully tested a selective laser melting–printed rocket injector that generated 20,000 pounds of thrust during a hot fire test,[8] proving that 3D-printed objects can accomplish heavy-duty tasks.

---

6   Dan Simmons, "Airbus Had 1,000 Parts 3D Printed to Meet Deadline," *BBC News*, May 6, 2015.

7   Yoav Zitun, "The 3D Printer Revolution Comes to the IAF," ynetnews.com, July 27, 2015.

8   GE Additive, "What Is Additive Manufacturing?" webpage, undated.

Potential pitfalls include limits on types of materials available and product sizes. Although 3D printing techniques are available for many classes of materials, including resin, ceramic, plastic, and metal, they might not be available in the specific material needed. Product sizes are largely limited by the size of the printer, which is crucial for on-board fabrication. Additionally, 3D printing is unsuited for large-scale manufacture. The larger the number of specific parts needed, the more relatively efficient other fabrication processes become. Also, the layer-by-layer nature of additive manufacturing can introduce structural weaknesses. Depending on the manufacturing process, there also might be significant post-processing steps (e.g., water jetting, sanding, chemical soak) and possible emissions hazards. Lastly, even with an on-board fabrication device, ships would still need to store raw materials for production.

Capability Development Summary

Several of these technologies might assist with the NSFS mission, but none are at the moment part of any POR. This is because, among other things, there is no stated requirement for the capability that these would address and thus no good way of assessing how much value the capabilities would add, given a certain cost and level of effort. However, some capabilities might have value under a wide variety of cases, particularly those that mitigate volume and capacity shortfalls.

When assessing capabilities that add range, it is important to separate the idea of sufficient stand-off range to avoid engagement by shore defenses—which is very unlikely to ever be achieved by any capability—from sufficient range to service multiple separated targets. Consequently, solutions need not seek the maximum range, and there are capabilities not readily available, such as the hypervelocity projectile, that might meet requirements identified in the scenarios. An added advantage is that this kind of projectile is particularly adaptable for technologies that provide lighter and stronger materials or allow local manufacture.

One of the technologies that could be most effective in addressing shortfalls is also one that is already well developed but faces policy constraints: cluster munitions. Although we are not questioning the importance of humanitarian considerations, cluster munitions should be considered for cases where volume is a military requirement and any munitions use is likely to result in collateral damage. The effort to minimize dud rates and delayed explosions would be an important part of preserving this option.

CHAPTER SEVEN

Conclusion, Findings, and Recommendations

Although NSFS is a mission with a long history—dating at least to the Civil War—it also is one that has not received an appropriate amount of scrutiny in terms of requirements development. The Navy equips its guided-missile cruisers and destroyers with guns, in accordance with ROC and POE requirements. Any other attempt to fulfill requirements has been a matter of a courteous attempt to respond to letters or infer informal requirements from operational concepts, none of which reaches the level of a formally stated performance parameter.

Although the need for better requirements development is a finding, it does not seem to be a matter that should require much debate or controversy. If USMC wants NSFS, it needs to formally specify what it needs. Until that is done, the Navy will continue on the path of meeting ROC/POE requirements and treating the remainder of capability development as merely a suggestion or a desire, largely because that is all that currently exists in terms of actual guidance.

Capability Findings

We have already mentioned that the requirement for NSFS is not developed and thus cannot be met other than in general provision of capability. However, our case studies and modeling did suggest the following:

- Targeting, particularly in denied environments, is likely to be very challenging and will be highly dependent on organic assets, principally UAVs. This issue is not confined to NSFS. Targeting in general is likely to be a challenge in many warfare areas. However, NSFS requirements for rapid and accurate firing information make the problem particularly difficult.

- Sensor-to-shooter timelines are far too long to support effective engagement on a fluid battlefield. NSFS for maneuvering forces ashore has to be capable of responding at very short notice to calls for fire.

- A single ship firing rounds from a single gun, even if targeting is optimal and C2 is well executed, is physically limited in the targets it can reach and the numbers of targets it can simultaneously service. This suggests that a single-ship model simply might be unworkable in heavily contested environments, no matter the capability and capacity of individual ships; however, this does not render the issue unimportant. Some autonomous vessel proposals could alleviate some of this shortfall, although none of the proposals necessarily result in platforms that would be capable of persistence and volume.

- The Navy has selected its mix of munitions based on considerations of what it believes to be its most likely operational mission, but as a result has undervalued magazine munitions that might be particularly valuable. In addition, its POR for munitions does not address area effects.

- Most of the scenarios we consider involve a considerable expenditure, and formal modeling against a plausible target set indicates a very high volume of munitions expenditures, generally beyond what would be carried in a ship's magazine. Achieving suppression as opposed to destruction lowers demand, as does use of area munitions, but the fundamental conclusion is that although an exact volume requirement is not defined, it is unlikely that the POR can even approach it. New technologies, such as 3D printing, that allow for on-site manufacture of ordnance could mitigate the magazine size shortfall.

Recommendations

The most obvious and most compelling recommendation is that USMC should identify what it needs from the Navy, using some combination of scenario and quantitative analysis. Absent a formal definition of requirements, the Navy has neither the incentive nor the reason to go beyond what is stated in the ship basis ROC/POE documents.

- Regardless of what requirements are ultimately ratified, USMC and USN should continue to invest in organic airborne ISR, which can be used even when parts of the electromagnetic spectrum are denied.
- Regardless of eventual requirements determination, USMC and USN should invest in tactical C2 solutions that allow compression of sensor-to-shooter timelines.
- Assuming requirements do get determined according to what seem to be likely scenarios, the following additional investments should be considered:
  - area munitions to challenge enemy maneuver capability
  - lighter munitions that allow extension of range, specifically to allow ships to service multiple landing force targets from a single location
  - ship modifications for larger magazines
  - unmanned fire support platforms that can be put into direct support roles
  - additive manufacturing to allow for production of gun ammunition to increase on-station time during periods of high use.

Bibliography

Alexander, Robert Michael, *An Analysis of Aggregated Effectiveness for Indirect Artillery Fire on Fixed Targets*, thesis, Atlanta, Ga.: Georgia Institute of Technology, 1977. As of January 23, 2020:
https://smartech.gatech.edu/bitstream/handle/1853/25669/alexander_robert_m_197708_ms_119723.pdf

Boeing, "V-22 Osprey," webpage, undated. As of January 23, 2020:
https://www.boeing.com/defense/v-22-osprey/

Bonds, Timothy M., Joel B. Predd, Timothy R. Heath, Michael S. Chase, Michael Johnson, Michael J. Lostumbo, James Bonomo, Muharrem Mane, and Paul S. Steinberg, *What Role Can Land-Based, Multi-Domain Anti-Access/Area Denial Forces Play in Deterring or Defeating Aggression?* Santa Monica, Calif.: RAND Corporation, RR-1820-A, 2017. As of May 16, 2019:
https://www.rand.org/pubs/research_reports/RR1820.html

Calloway, Audra, "Army Developing Safer, Extended Range Rocket-Assisted Artillery Round," Picatinny Arsenal Public Affairs, August 25, 2016.

Center for Naval Analyses, *Naval Surface Fire Support AoA Final Report*, Arlington, Va., November 2009.

Chairman of the Joint Chiefs of Staff, *Countering Air and Missile Threats*, JP 3-01, Washington, D.C.: Joint Staff, April 21, 2017.

Chief of Naval Operations, Surface Warfare (N96), *Required Operational Capabilities and Projected Operational Environment for (Arleigh Burke) Class Guided Missile Destroyers*, OPNAV F3501.311C, July 28, 2017.

Cluster Munition Coalition, "Use of Cluster Bombs," webpage, undated. As of January 23, 2020:
http://www.stopclustermunitions.org/en-gb/cluster-bombs/use-of-cluster-bombs/a-timeline-of-cluster-bomb-use.aspx

De La Fuente, Jesus, "Graphene, What is It?," web page, undated. As of December 10, 2019:
https://www.graphenea.com/pages/graphcnc

Di Falco, Andrea, Martin Ploschner, and Thomas F. Krauss, "Flexible Metamaterials at Visible Wavelengths," *New Journal of Physics*, Vol. 12, November 2010. As of March 25, 2020:
https://iopscience.iop.org/article/10.1088/1367-2630/12/11/113006

Duplessis, Brian, "Thunder from the Sea: Naval Surface Fire Support," *Fires*, May-June 2018.

Eckstein, Megan, "New Requirements for DDG-1000 Focus on Surface Strike," *USNI News*, December 4, 2017.

GAO—*See* U.S. Government Accountability Office.

GE Additive, "What Is Additive Manufacturing?" webpage, undated. As of January 24, 2020:
https://www.ge.com/additive/additive-manufacturing

Gertler, Jeremiah, *Operation Odyssey Dawn (Libya): Background and Issues for Congress*, Washington, D.C.: Congressional Research Service, March 30, 2011.

Guarino, Ben, "The Navy Called USS Zumwalt a Warship Batman Would Drive. But at $800,000 per Round, Its Ammo Is Too Pricey to Fire," *Washington Post*, November 8, 2016.

Hanlon, Edward, Jr., "Naval Surface Fire Support Requirements for Expeditionary Maneuver Warfare," memorandum to Chief of Naval Operations (N7), Quantico, Va., March 19, 2002.

Hastrich, Carl, "Self Healing Materials," *Bouncing Ideas*, February 1, 2012.

IHS Markit, "Jane's Sentinel Security Assessment–China and Northeast Asia," webpage, February 13, 2019.

Isely, Jeter A., and Philip A. Crowl, *The U.S. Marines and Amphibious War: Its Theory, and Its Practice in the Pacific*, Princeton: Princeton University Press, 1951.

Joint Requirements Oversight Council, *Initial Capabilities Document for Joint Fires in Support of Expeditionary Operations in the Littorals*, Washington, D.C.: U.S. Joint Chiefs of Staff, JROCM 274-05, November 1, 2005.

Kingery, Ken, "Beyond Materials: From Invisibility Cloaks to Satellite Communications," *DukeStories*, March 25, 2018. As of February 5, 2020:
https://stories.duke.edu/beyond-materials-from-invisibility-cloaks-to-satellite-communications

Lamothe, Dan, "Missiles from Rebel-Held Yemen Fired at U.S. Navy Destroyer, Saudi Military Base," *Washington Post*, October 10, 2016. As of January 22, 2020:
https://www.washingtonpost.com/news/checkpoint/wp/2016/10/10/missiles-from-rebel-held-yemen-territory-fired-at-u-s-navy-destroyer/?utm_term=.b54842da704e

Lappin, Yaakov, "IAI Announces New Mini Harpy Loitering Munition," *Jane's Defence Weekly*, February 19, 2019.

Marine Corps Combat Development Command, Fires and Maneuver Integration Division, *Naval Surface Fire Support Requirements and Program Updates*, information paper, Quantico, Va., May 1, 2018.

MCCDC—*See* Marine Corps Combat Development Command.

Missile Defense Advocacy Alliance, "OTR-21 Tochka (SS-21 Scarab)," webpage, 2019. As of January 23, 2020:
https://missiledefenseadvocacy.org/missile-threat-and-proliferation/
missile-proliferation/russia/ss-21-mod-2/

Naval Vessel Register, "TICONDEROGA (CG 47)," webpage, December 12, 2017. As of January 22, 2020:
https://www.nvr.navy.mil/SHIPDETAILS/SHIPSDETAIL_CG_47.HTML

———, "USS Arleigh Burke (DDG 51)," webpage, June 27, 2018. As of January 22, 2020:
https://www.nvr.navy.mil/SHIPDETAILS/SHIPSDETAIL_DDG_51.HTML

———, "No Name (DDG 138)," webpage, September 27, 2018. As of January 22, 2020:
https://www.nvr.navy.mil/SHIPDETAILS/SHIPSDETAIL_DDG_138.HTML

———, "USS Paul Ignatius (DDG 117)," webpage, June 7, 2019. As of January 22, 2020:
https://www.nvr.navy.mil/SHIPDETAILS/SHIPSDETAIL_DDG_117.HTML

North Atlantic Treaty Organization, *NATO Glossary of Terms and Definitions (English and French)*, AAP-06, Brussels, Belgium, 2019.

Office of the Secretary of Defense, *Annual Report to Congress: Military and Security Developments Involving the People's Republic of China 2018*, Washington, D.C.: U.S. Department of Defense, May 16, 2018. As of September 24, 2019:
https://media.defense.gov/2018/Aug/16/2001955282/-1/-1/1/2018-CHINA-MILITARY-POWER-REPORT.PDF

O'Rourke, Ronald, *Navy Lasers, Railgun, and Hypervelocity Projectile: Background and Issues for Congress*, Washington, D.C.: Congressional Research Service, October 21, 2016.

———, *Navy DDG-51 and DDG-1000 Destroyer Programs: Background and Issues for Congress*, Washington, D.C.: Congressional Research Service, last updated December 17, 2019.

Sherwood, John Darrell, *War in the Shallows: U.S. Navy Coastal and Riverine Warfare in Vietnam, 1965–1968*, Washington, D.C.: Naval History and Heritage Command, Department of the Navy, 2015.

Simmons, Dan, "Airbus Had 1,000 Parts 3D Printed to Meet Deadline," BBC News, May 6, 2015. As of January 24, 2020:
https://www.bbc.com/news/technology-32597809

South, Todd, "Plan Would Double Artillery Upgrades in Army Arsenal over the Next Five Years," *Army Times*, March 22, 2019. As of January 22, 2020:
https://www.armytimes.com/news/your-army/2019/03/22/
plan-would-double-artillery-upgrades-in-army-arsenal-over-the-next-five-years/

Telford, Mark, "The Case for Bulk Metallic Glass," *Materials Today*, Vol. 7, No. 3, March 2004, pp. 36–43.

U.S. Department of the Navy, *U.S. Navy Program Guide 2017*, Washington, D.C., undated.

U.S. Government Accountability Office, *Defense Acquisitions: Challenges Remain in Developing Capabilities for Naval Surface Fire Support*, Washington, D.C., GAO-07-115, November 2006.

———, *Weapon Systems Annual Assessment: Limited Use of Knowledge-Based Practices Continues to Undercut DOD's Investments*, GAO-19-336SP, Washington, D.C., May 2019. As of January 22, 2020:
https://www.gao.gov/products/GAO-19-336SP

U.S. Joint Chiefs of Staff, *Department of Defense Dictionary of Military and Associated Terms*, Joint Publication 1-02 (JP 1-02), Washington, D.C., November 8, 2010.

U.S. Marine Corps, *Fire Support Coordination in the Ground Combat Element*, MCTP 3-10F, Washington, D.C., April 4, 2018.

USMC—*See* U.S. Marine Corps.

USN—*See* U.S. Navy.

U.S. Navy, "Fact File: Cruisers – CG," webpage, last updated January 9, 2017. As of February 5, 2020:
https://www.navy.mil/navydata/fact_display.asp?cid=4200&tid=800&ct=4

U.S. Navy, "Fact File: MK 45 – 5-Inch 54/62 Caliber Guns," webpage, last updated January 16, 2019a. As of January 22, 2020:
https://www.navy.mil/navydata/fact_display.asp?cid=2100&tid=575&ct=2

U.S. Navy, "Fact File: Destroyers – DDG," webpage, last updated August 21, 2019b. As of February 5, 2020:
https://www.navy.mil/navydata/fact_display.asp?cid=4200&tid=900&ct=4

Welch, Shawn A., *Joint and Interdependent Requirements: A Case Study in Solving the Naval Surface Fire Support Capabilities Gap*, thesis, Norfolk, Va.: Joint Advanced Warfighting School, 2007. As of January 23, 2020:
https://pdfs.semanticscholar.org/a130/9f9efac7d0a81cbcb9e6966e0749788f98b2.pdf?_ga=2.75276306.1866012005.1579785395-952500926.1571234735

Weller, Donald M., "The Development of Naval Gunfire Support in World War Two," in Merrill Bartlett, ed., *Assault from the Sea: Essays on the History of Amphibious Warfare*, Annapolis, Md.: Naval Institute Press, 1983.

Zitun, Yoav, "The 3D Printer Revolution Comes to the IAF," ynetnews.com, July 27, 2015. As of January 24, 2020:
https://www.ynetnews.com/articles/0,7340,L-4684682,00.html